The Unitary Theory

Ulisse Di Corpo & Antonella Vannini

www.sintropia.it

Copyright © 2016 Ulisse Di Corpo & Antonella Vannini

ISBN: **9781520237527**

CONTENTS

Acknowledgments

Prologue   1

Luigi Fantappiè   3

Ulisse Di Corpo   33

Antonella Vannini   59

Epilogue   75

ACKNOWLEDGMENTS

This book was inspired by the work of professor Luigi Fantappiè and his 1942 *"Unitary Theory of the Physical and Biological World."*

We want to express our gratitude to Elena Fantappiè for her precious help and collaboration.

PROLOGUE

Few days before Christmas 1941, while talking with a physicist and a biologist, the mathematician Luigi Fantappiè had the insight of the Unitary Theory. In the 1920s physicists had rejected half of the solutions of the fundamental equations of the universe, since they imply retrocausality and the possibility of perpetual motion. Analyzing the properties of the solutions that had been discarded Fantappiè noticed that they explain the mysteries of biology, such as energy concentration, the increase in differentiation, complexity and order, whereas the solutions which had been accepted are governed by the law of entropy. Fantappiè coined the term syntropy (combining the Greek words *syn*=converging and *tropos*=tendency).

In that same period other scientists had similar insights. Among them the paleontologist Pierre Teilhard de Chardin and the psychiatrist Wilhelm Reich. Teilhard de Chardin died on April 10, 1955, his books were removed from all the book sellers and libraries and the Vatican emitted a decree ordering the withdrawal from circulation of the works of Teilhard, together with all those books which favored this erroneous doctrine. Luigi Fantappiè died on July 28, 1956, his Unitary Theory immediately disappeared from libraries and became unavailable; from his private archive all the envelops dealing with syntropy and the

Unitary Theory were removed. Wilhelm Reich died on November 3, 1957, all his books and articles were burned, probably the worst example of censorship in U.S. history.

On the 19th of April 1977 I myself had the insight of syntropy and the Unitary Theory. One major difference of my work is that I start from the dual-time solution of Einstein's energy/momentum/mass equation, whereas Fantappiè starts from the dual-time solution of the D'Alembert operator and consequently expects syntropic phenomena not only in quantum mechanics but also in electromagnetism.

In 2001 Antonella Vannini provided impressive experimental evidence to this theory. The hypothesis Antonella was working on is very simple: "*If life is sustained by syntropy, the systems which sustain life, such as the autonomic nervous system, should show retrocausal activations.*" Several researchers had already found that the heart rate and skin conductance react in advance to stimuli.

In the form of an imaginary conference this book allows Luigi Fantappiè to present his Unitary Theory, followed by Ulisse Di Corpo and Antonella Vannini's contributions and extensions.

Ulisse Di Corpo
Rome, 21 December 2016

LUIGI FANTAPPIE'

Thank you for the opportunity that you are providing to present again, after 75 years, my *Unitary Theory of the Physical and Biological World*.

I am a mathematician. I was born in Viterbo, Italy, on September 15, 1901. I graduated from the most exclusive Italian university, the Scuola Normale Superiore di Pisa, at the age of 21. During the University years I became good friend with Enrico Fermi. I was very well known among physicists, and after my dissertation I spent some time in Paris and in Germany giving lectures. I came back to Italy, with an academic position assigned at the University of Rome where I became full professor. Before the Second World War in the years 1934-1939 I was in Brazil, San Paolo, where I founded the faculty of mathematics. In April 1951 Oppenheimer invited me to become a member of the exclusive Institute for Advanced Study in Princeton and work directly with Einstein.

I died very young during the night between the 28th and 29th of July 1956. I had a house in Bagnaia, a beautiful medieval town near Viterbo. I was there for the weekend. On Sunday afternoon I should have gone back to Rome. One of my students was going to discuss his final dissertation on Monday morning. On Saturday afternoon I was feeling fine, I spent the entire afternoon trekking in the hills, with my faithful little dog. That evening two persons came to visit me. They wanted to talk about their intuitions. That night I suddenly had a heart failure and died.

You might find it strange that a mathematician has adventured himself in such a wide exploration in the fields of other sciences, without having a specific knowledge of them. This consideration stopped me for a long time in letting this theory become public. But when I outlined its content to my friend and colleague Professor Azzi of the University of Perugia and having received from him a strong and positive support, I felt I had to formulate it in a more detailed way and discuss it with colleagues of other disciplines. The encouragement which I received convinced me that this theory offers a unique possibility towards a Unitary Theory in which all the phenomena which we know naturally take place. It allows to treat within the same rational frame the physical, chemical and biological phenomena, including also those of consciousness and personality. It also provides interpretations of the fundamental phenomena of quantum mechanics.

It was in the days just before Christmas 1941, as a consequence of conversations with two colleagues, a physicist and a biologist, that I was suddenly projected in a new panorama, which radically changed the vision of science and of the Universe which I had inherited from my teachers, and which I had always considered the strong and certain ground on which to base my scientific investigations.

Suddenly I saw the possibility of interpreting a wide range of solutions, the *advanced potentials* of the wave equation which can be considered the fundamental law of the Universe. These solutions had been always rejected as impossible, but suddenly they appeared possible, and they explained a new category of phenomena which I later named *syntropic*, totally different from the mechanical, physical and chemical laws, which obey only the principle of causation and the law of entropy.

Syntropic phenomena, which are represented by those strange solutions of the advanced potentials, obey two opposite principles of finality and differentiation and they are not causable in a laboratory.

Its finalistic properties justify the refusal among scientists, who accepted without any doubt the assumption that finalism is a metaphysical principle, outside Science and Nature. This assumption obstructed the way to a calm investigation of the real existence of this second type of phenomena; an investigation which I accepted to carry out, even though I felt as if I were falling in an abyss, with incredible consequences and conclusions.

It suddenly seemed as if the sky were falling apart, or at least the certainties on which mechanical science had based its assumptions. It appeared clear to me that these "syntropic", finalistic phenomena which lead to differentiation and could not be reproduced in a laboratory, were real, and existed in nature, as I could recognize them in the living systems.

The properties of syntropy opened consequences which were just incredible and which could deeply change the biological, medical, psychological, and social sciences.

I presented this theory for the first time on November 3, 1942, in Spain, at a conference at the *Consejo Nacional de Investigaciones Cientificas*. I then was invited to Barcelona by the *Academy of Science*, where on December 1, 1942, I discussed the details of the Theory in a private meeting.

On the days that go from the 31st of May to the 2nd of June 1943 I was invited by Prof. Carlini to the Science and Philosophy conference which was held at the *Scuola Normale Superiore di Pisa*. In this occasion I presented the Unitary Theory among scientists of the most diverse orientations. I was able to discuss the Unitary Theory with many prestigious colleagues, among whom professors Severi, Rondoni, Carrelli, Puccianti, Persico, Guzzo, Abbagnano and Banfi. I was given an entire afternoon for questions and answers. It was then that I decided to write *The Unitary Theory of the Physical and Biological World*.

The Unitary Theory:

- confirms the law of causality and the second principle of thermodynamics for all the phenomena which we call entropic. Causality, which until now was a conceptual category, becomes a law of the entropic phenomena, which has a precise and objective meaning.
- describes phenomena totally different from the entropic ones, which we can find in the mysterious properties of life. These phenomena are predicted and explained by the same equation which govern the entropic phenomena, but are essentially different and allow to see an immense panorama, which might be more vast, diversified and meaningful of the entropic phenomena.
- shows that the same wave equation which combines special relativity with quantum mechanics predicts syntropic and entropic phenomena. Syntropic phenomena are moved by attractors, finalities, whereas entropic phenomena are moved by causes.

Scientists have accepted, without providing any clear postulation that using the principle of causality all natural phenomena can be reproduced. The Unitary Theory shows that only the entropic phenomena can be caused and reproduced, whereas syntropic phenomena cannot be caused and reproduced, they can only be observed.

All the knowledge that has been produced in the last centuries using the experimental method, on which science is based, is limited to the entropic side of nature, whereas for the syntropic phenomena we need a new scientific methodology.

Syntropic phenomena can be influenced indirectly from specific entropic phenomena, but on the whole they constitute an extremely important part of the universe which is beyond our possibility of manipulation.

The entropic side of reality will inevitably fail to account for the totality, since the laws of nature are symmetric in regard to time and can be diverging/entropic and converging/syntropic, and this last type of phenomena are those which are at the essence of my discovery.

If we look at the present knowledge of the intimate structure of the Universe, we see that it can be summarized in three basic points:

- Dalton's atomic theory established in the XVIII century and later improved by Stanislao Cannizzaro, with the distinction of molecules and atoms, and then by Lorentz who formulated the particle theory of electromagnetism and Planck and Einstein with the quantum theory of energy. These results on the intimate atomic-particle nature of matter of the entire Universe is now considered acquired, since it has been

tested and validated for more than two centuries.
- The wave nature of all the physical phenomena, when considered in their most profound essence, at the level of quantum mechanics. Studied by Heisenberg, Schrödinger, Dirac, and others, has given birth to modern nuclear physics. The wave nature of the physical phenomena can now be considered acquired thanks to the experimental validation of Davison and Germer with electron rays which shows diffraction and interference properties in particles. These properties are typical of waves.
- The validity of the Theory of Special Relativity, which has received corroboration at the atomic level, such as the explanation of the increase in mass, the inertia of the electron, and the increase in speed. This theory leads to a description based on four dimensions which unites space with time, reaching in this way a perfect symmetry among the spatial and time dimension. This representation is named chronotope.

Let us now see how these three fundamental elements can be harmonized.

First of all the atomic-particle nature of matter and the wave manifestation seemed to conflict, since one is deterministic and the other probabilistic.

At the moment this conflict has been solved saying that it is impossible to predict in a deterministic way the behavior of particles since the prediction is attributed to waves which are probabilistic. Waves offer a deterministic prediction only when we consider large numbers of particles.[1]

In Boltzmann and Poincaré theory the Universe was described as governed by strictly deterministic laws, both at the macro and at the micro level, where probability was used in a way which was considered only to be temporary, with the belief that the evolution of science would have replaced the mean values of probability with the exact values of the rigorous deterministic laws, which were believed to be at the foundation also of the microcosm.

Now, instead, the probabilistic laws of these phenomena are considered to be at the foundation of the Universe, whereas the deterministic laws, which are valid at the macrocosm level, are considered to be only a consequence of the law of large numbers.

In 1927 Schrödinger renounced to special relativity in the formulation of his wave equation since in quantum mechanics waves should propagate at infinite speeds, and this is in conflict with the theory of special relativity which prohibits speeds greater than the speed of light.[2] The conflict between Schrödinger's non-relativistic wave equation and special relativity is obvious also at the general level, since time appears in a non-symmetric way, as a first derivative.

It is generally accepted that Schrödinger's wave equation is only a temporary description of the quantum phenomena, which is valid with good approximation only in those cases in which the speed of light can be considered infinite, but which will have to be replaced by a quantum-wave theory which is more exact and agrees with special relativity.

On the contrary relativistic wave equations are symmetrical for all four variables, the space variables x, y, z and the time variable t, in agreement with special relativity. In this way a second order equation is obtained not only for the space variable, but also for time, and the D'Alembert operator is used.

The study of such an equation was brilliantly conducted by Dirac, considering all its implication, in the case of the electron, decomposing the equation of the second order in an equation of the first order, and showing that this wave-relativistic equation of the electron allowed the full explanation of phenomena that until then where difficult to understand rationally, such as the magnetic momentum of the electron, which we now call the spin, which is due to the rotation of the electron on itself. Dirac found in his equation that beside the usual electron, also a symmetrical solution appeared, a neg-electron which is now named positron, which had not been observed since then and which was considered to be impossible. But after a short time, the positron was discovered by Blackett and Occhialini, and this validated the prediction that Dirac's equation made of this particle, showing at the same time the strong foundation of quantum mechanics when combined with special relativity.[3]

It is important to underline that although we don't yet have the details of the partial derivatives equations which describe in all their details the various quantum systems, we can determine some very important characteristics of these unknown differential equations, such as the fact that the properties of the characteristic cone will apply to all, and the fields of dependency and influence of the solutions, which are described by Dirac's equation.

These properties have been deduced from those of the D'Alembert operator, which is linked only to the geometrical nature of the chronotope, and does not depend from the particular properties of the particle, which are instead described by the other terms of the equation which do not influence at all the geometrical nature of the chronotope. The chronotope does not vary when we consider a different type of particle, or particle systems, we will have that also for the equations of unknown partial derivatives, which support these quantum systems, the characteristic cone and the fields of dependency and influence of the solutions will be the same of those that Dirac found in his equations.[4]

The fundamental solutions of the D'Alembert operator have been provided by Poincaré[5], Ritz[6] and Giorgi[7]. A first solution describes waves diverging from the source and are named *delayed potentials*.[8] A second solution describes waves converging to the source and are named *advanced potentials*.

The criticisms to the possibility of advanced waves were made mainly by Wiechert, Lorenz, Poincaré, Ritz and Giorgi, who considered that if converging waves existed it would be possible to concentrate energy and in this way to devise a perpetual motion machine. And this was considered to be impossible.

Now, let us see how the notion of cause and causality, as they are understood by physicists and modern scientists, differ from the more general "deterministic principle", considered as the possibility of making a prediction.

When we say that the event A causes B, we believe that once we have observed A we can certainly predict that B will become true. But, we can also predict that after the event of night the Sun will rise, however no one can say that the rise of the Sun is caused by the night. In the notion of causality there is something more.

When can we say that A causes B?

The answer to this question must be searched in the experimental method, which Galileo put at the foundation of all the modern sciences.[9]

A is the cause of B when we insert experimentally A and we observe B.

But in order to have a convincing experiment we need to be free, at least within certain boundaries to cause A, where and when we wish. As a matter of fact if someone would want to convince me that A is the cause of B producing A in order to asses B, only in a specific place and time, I would remain skeptic.

The experimental method provides an exhaustive answer to the question if A is the cause of B, only when we have the total freedom to produce A and see if B follows. Only in this condition we can be sure that A is the cause of B. This leads to the important conclusion that we can recognize the events which are the cause of others only thanks to the free will of the experimenter.

Causality gives way to the more general and objective "determinism" which tries to determine past and future events analyzing present events. But also determinism has shown to be insufficient in the study of particles, leaving the field to a wider perspective in the microcosm, which is based on probability.

We can state that widening our knowledge the categories which we were trying to apply have widened, moving from the law of causality, to determinism, to the modern probabilistic theories of quantum mechanics.

What I have just said does not mean that causality and determinism should be abandoned; they cannot be used to explain all the reality.

Causality and determinism are certainly useful and fundamental in the study of a well-defined part of reality. When we move from wave mechanics to the more limited deterministic field of the macrocosm, where the law of large numbers applies, probabilities change into frequencies which can be handled in a deterministic way.

If we isolate the system in such a way that nothing happens beside what the experimenter wants with his free-will and B is different from zero only from the moment when A is produced, we can state that A causes B. The cause becomes the source which causes B and, therefore, each event B which is caused by A, is always affected by diverging waves from the point A. The solution that governs B will therefore be of the type of the *delayed potentials*.

What I said implies that causable phenomena are always entropic. Each entropic phenomenon, each phenomenon based on diverging waves has its cause in the source which causes the diverging waves.

In this way we get to the fundamental theorem: *a necessary and sufficient condition for B to be entropic, is that it can be caused using another phenomenon A, which is the source from which the diverging waves that constitute B are emitted.*

The majority of the physical and chemical phenomena, which we can study in our laboratories, are entropic. Causality applies to entropic phenomena, such as those studied in mechanics, acoustics, optics, electromagnetism and chemistry. This does not exclude that in nature we can have other phenomena, beside the entropic ones, such as the syntropic phenomena, which cannot be caused using our free-will, since they would then fall within the entropic phenomena.

Diverging waves imply necessarily the second law of thermodynamics, which states that entropy does not diminish, but increases during time.

From an intuitive point of view we can consider entropy as a state of leveling of a large number of particles. Diverging waves dilute in spaces which are always bigger, and if the space is limited, as it happens in a container, their intensity tends to level.

The wave equation extends this law to all the phenomena which are governed by diverging waves and in this way the second law of thermodynamics is no longer obtained from a probabilistic postulate, such as Clausius' principle of the elementary disorder, but it is a logical and necessary consequence of the law of causality. When the law of causality applies to a phenomenon, we can say that this phenomenon is entropic.

This is the reason why it is impossible to obtain a perpetual motion machine. The degradation of energy is a necessary and logical consequence of the law of entropy which applies to all the machines. The main argumentation which is used in order to exclude advanced potentials is that they would allow to devise perpetual motion machines, converging the energy that was first dispersed towards a point and then diverging it, then again converging it, and so on forever.

The main characteristics and properties of those phenomena which are constituted by advanced waves, which I have named syntropic, are profoundly different from the entropic phenomena previously described:

— They *cannot be caused* by our free will, at least in their essential components constituted by the converging waves, since on the contrary they would fall in the category of the entropic phenomena, which are governed by the law of causality, and characterized by diverging waves. For the same reason, syntropic phenomena can be influenced, in their evolution, only indirectly by specific entropic phenomena, the only which we can use, which can interfere with them, for example by modifying the environment in which they take place, since it is plausible that if the two phenomena exist they are not separated in nature, but intertwined.

- They *concentrate energy* within always smaller spaces. Also the particles represented by these waves progressively concentrate in the center of the waves. Whereas the entropic systems go from concentrated to dispersed, in the syntropic phenomena exactly the opposite happens. We first have dispersed phenomena which concentrate in always smaller spaces. The entropic phenomena manifest with dissipative characteristics. An example is when we light a match. We have a cause which is concentrated in a small space, from which the light irradiates, with an intensity that diminishes with the distance, diluting the effect. Syntropic phenomena manifest with an anti-dispersive character, a converging manifestation, which goes from diluted to concentrated in specific points. Whereas the entropic phenomena radiate from specific points, syntropic phenomena concentrate towards specific points.

- The *concentration of energy cannot be endless*. Since it cannot continue indefinitely, after a period of syntropic concentration entropic dissipation takes over. This means that we witness a process of exchange of matter and energy. Incoming energy and matter indicate syntropic processes, outgoing energy and matter indicate compensatory entropic processes.

- *Entropy diminishes*, since with time differentiation

increases. From a rigorous formal point of view syntropy has the same value of the second law of thermodynamics.
- We see a *tendency towards differentiation and complexity*. Syntropic phenomena show in complex forms, as it happens with biological systems which cannot be explained in a satisfactory way by using only their physical and chemical properties.
- They are *in a continuous state of energy dissipation* (warm bodies), and this is a consequence of the fact that syntropic systems absorb energy but they don't evolve towards heat death.

I suggest that it is possible to scientifically study syntropic phenomena considering that the D'Alembert equation is time reversal. This equation is symmetrical in respect to time. Reversing the time variable all the solutions of the delayed potentials become solutions of the advanced potential, and vice versa. Consequently, a very simple way to obtain the syntropic properties of a system from the entropic ones is just to invert the time direction.

Nearly all the phenomena are dual phenomena. In our language this is usually expressed by adding the prefix "anti": combustion becomes anti-combustion, filtration anti-filtration, matter anti-matter, energy anti-energy, etc… Applying this principle of duality we can obtain the characteristics of the syntropic phenomena from its dual entropic phenomena.

According to the D'Alembert equation, entropic phenomena are activated when waves start diverging from the source. For example when we light a match electromagnetic waves start diverging at the speed of light in all the directions in a uniform way.

When we reverse the flow of time the dual syntropic phenomena shows. Waves concentrate towards the center of the sphere, increasing their intensity. These waves would be uniformly distributed in all the directions, independently from where they seem to come.

Let us consider the waves which propagate on a pond. We can cause this phenomenon, which is therefore entropic, by throwing a stone in the pond and observe how the waves propagate and diverge. The dual syntropic phenomenon would show these waves perturbations concentrate in a point from which the stone would then emerge, leaving behind the water at rest. If we could observe such a phenomenon we would think that some sort of intelligent being had organized it.

Now, let us imagine a brand new telescope that we have forgotten in our garden. At first rust forms, then it falls and breaks into pieces. Pieces of metal and glass gradually deteriorate and mix with the ground. Changing the time flow we would see that from the ground different pieces of metal and glass separate, then they find their place in a design of lenses and tubes which form the telescope until a brand new and perfectly functioning telescope is reached. What puzzles us is the finalistic aim, which we usually attribute to the action of an intelligent being. Syntropic processes express finality, a purpose, intelligence as if a will is acting on them.

Finality is the characteristic of the syntropic phenomenon.

The law of causality and the law of finality are logical consequences of the intimate duality of the fundamental laws of physics. It is possible to state that without causes entropic phenomena cannot exist and without finalities syntropic phenomena cannot exist. Without causes and finalities the wave equations would be null. Consequently finality is not an accidental manifestation in a syntropic phenomenon, but it is a necessary condition of the syntropic phenomenon, without which it could not exist.

Science has investigated the entropic physical and chemical characteristics of life, without grabbing the essence of life. It is now well acquired in biology, thanks to the experiments devised by Pasteur, that there is no possibility of spontaneously producing life without starting from a minimum amount of life. This is referred to using the Latin words «vivum nisi ex vivo». Life stems from life. It is impossible to create life at our will. The non-causability of life tells that it is a syntropic phenomenon. It is also well known that vital phenomena cannot be influenced directly, but only indirectly. For example we cannot produce directly a plant or an animal with our hands, but we can only grow or raise them.

All living organisms concentrate in their body matter and energy. This tendency is visible especially in plants and it is due to the chlorophyllian process. We can therefore assume that in plants there is a quantitative prevalence of the converging syntropic phenomenon, which is also present in animals in their growth stage and then it is balanced with entropic processes at the adult stage, which start becoming gradually more relevant with aging and then totally prevailing with death. It is interesting to note that in metabolism the syntropic processes of absorption of matter and energy and construction of structures is names *anabolic*, whereas the entropic processes of dissipation, destruction of structure and release of energy and matter are named *catabolic*.

The syntropic process of energy absorption is always coupled with its dual phenomenon of energy dissipation. One of the major properties of life is that it is constantly releasing energy. This constant release of energy and by-products is coupled with the assimilation of matter and energy. A process of exchange of matter and energy which is named metabolism.

During the growth period, anabolic processes are

prevalent and an increase in differentiation is observed.

It is interesting to note that the probability that the smallest protein molecule arises by chance is less than 10^{-600}. This is an incredibly small number, represented by a 0 followed by 600 zeros and at the end, on the right, the number 1. In other words, the spontaneous formation of the smallest life molecule results to be practically impossible. The incredible amount of proteins that life shows conflicts with the second law of thermodynamics. This means that the law of entropy does not apply to life and that life is not an entropic phenomenon.

Finality is the fundamental characteristic of any syntropic phenomena, similarly to the principle of causality which is the fundamental characteristic of any entropic phenomena.

Only thanks to the principle of finality we can logically understand the smallest and most complex architecture of the living systems. Organisms differentiate in organs which are harmonically coordinated and arranged in order to reach a purpose. For example, the development of the eye starts from cells which are very similar, which then differentiate and take place in such ways that they build the elements of a perfect eye, such as lenses, vitreous body, which are by far more complex of a single protein.

The principle of finality shows that pretending to understand life through its physical and chemical elements, which are governed by causality, is just an illusion. Finality on which life is founded is similar and dual to the principle of causality which governs the entropic systems. Causality is the essence of the physical world, finality is the essence of life. Living systems tend towards aims and purposes. Life systems have a mission, and the greater the mission is, the more complex is the living system, with complex organs meant to reach its purpose.

The difficulty with the principle of finality is commonly found in the various theories of evolution. If we examine the most popular one, Darwin's theory of evolution, we see that it is based on three facts: the variability of life forms, the fight for survival, and the long permanence of life on Earth. These facts cannot be denied, but are not sufficient to explain life and all the various species of organisms.

In 1865 Mendel's experiments on plant hybridization seemed to prove the theory of evolution which Charles Darwin had published in 1859. But with Mendel we are not witnessing the formation of new species, we are witnessing the separation of genetic information into different characters and forms.

According to Darwin at the beginning on Earth only few simple unicellular life systems could exist.

Darwin introduces the concept of random variability as the cause of new species. About randomness, the probability of the random formation of any living system can be calculated using the kinetic theory of gasses which considers all the possible combinations with the same probability. Using this assumption the probability of the formation of the smallest protein is less than 10^{-600}. It is therefore easy to imagine how smaller the probability of the formation of an organ is, such as the eye, the ear, or any of the apparatuses that we commonly use. The probability of the formation of a whole animal is even smaller. The random permutations which are required for the formation of just one protein are greater than all the possible permutation in the history of the entire Universe. Consequently, the long permanence of life on Earth is insufficient to account for the formation of the smallest forms of life and of any living being. The probability of life happening by chance are by far smaller than the probability of witnessing water freezing when put in a pot placed on the flame of a cooker.

And, if life is caused it should obey the law of entropy and go towards the dissolution of any form of organization and complexity. With time we would see the increase of entropy and it is illogical to pretend that complexity can be achieved at the expenses of other beings or using the light of the Sun since in the first stages of the evolution of life on Earth, there weren't other beings and the atmosphere did not allow Sun rays to reach the land.

When on the contrary we consider life as a syntropic

phenomenon, the principle of finality applies and leads to increase differentiation, complexity and harmony.

The planet Earth can be considered as an immense living organism. The fact that species are interdependent, that they cannot live without others, for example fruits need insects for the pollination, we need vegetables ... all these species can be considered as parts of a more complex organism orchestrated by a finality, which can be reached only through differentiation.

In human beings cells cooperate towards greater ends and only in pathological situations, when they lose their end, they develop in an excessive way, suffocating other cells, as it happens with cancer.

At the beginning of evolution simple forms of life are the aim, then they become the foundation blocks for always higher forms of life. Species are not caused by previous species, but they are attracted towards future designs and forms.

Syntropy solves the profound dissymmetry that the second law of thermodynamics has introduced in the universe, by considering all the solutions of the fundamental equations. The theory of syntropy shows that the solutions that physicists wanted to exclude represent exactly the essence of life phenomena, that seemed impossible to be explained.

Syntropy is capable of unifying different scientific disciplines in a harmonic way, opening in this way the road to a unified theory, a theory of everything that encompasses in a coherent theoretical framework all the manifestation of the universe.

With the formulation of the experimental method the problem of science was considered definitely solved. This method considers causality at the foundation of all natural phenomena.

The experimental method is used to test cause and effect relations. In the case of positive results the hypothesis is accepted, otherwise it is rejected. Experiments provide the verdict which allows to separate what is true from what is false.

The experimental method is profoundly different from the method which Aristotle suggested, which was useful in the formulation of theories but did not provide a way to choose among the various hypotheses.

The experimental method implies the law of causality and has limited scientific investigation to entropic phenomena. We can therefore call the Galilean science an entropic science.

Let us analyze the experimental method. It is divided in

three steps: observation, formulation of a theory, experimental validations of its hypotheses.

As we have previously seen each entropic phenomenon has a dual syntropic phenomenon and vice versa. Consequently, although it is impossible to use the experimental method to test directly a syntropic hypothesis, we can set up an experiment in order to test the dual entropic hypothesis. In this way the study of the syntropic phenomena can be done indirectly studying the dual entropic phenomena.

Syntropic scientists would therefore have to search for the dual entropic phenomena, since when they manage to do this it is possible to progress using the experimental method.

Let us apply this dual method to a phenomenon which has yet to be explained, such as the absorption of water and nutrients from the land and their rise in the higher parts of the plant.

The hypothesis of osmosis does not stand since plants also acquire salts from the land. The idea that capillary conducts are responsible for the rise of water also does not stand when we consider that some trees can reach the height of 150 meters. These phenomena of absorption of water and rise of water seem to contradict the entropic laws of physics and this suggests that we are in front of syntropic phenomena which cannot be caused artificially. We can therefore apply to them the method of "dual experimentation".

In order to obtain the dual entropic phenomenon let us imagine that time flows in the opposite direction. We would see the lymph flow down until it reaches the roots and then water and salts disperse in the land. This dual image can be reproduced, for example, putting a non-living pole in the land and observing how water and salts filtrate from the top to the bottom and through the land. This entropic process of filtration, which can be easily caused in any moment proves that the process which we are witnessing in plants is the dual process of filtration. We can therefore name it anti-filtration.

One may object that in filtration gravity helps the process. Well, when we change the direction of time also gravity changes and from an attractive force it becomes a diverging repulsive force which helps water rise in the anti-filtration process which we observe in plants.

Now, let us take the combustion of vegetal tissues. This is a phenomenon which we can cause at our will and which is therefore certainly entropic. We see at the beginning a highly differentiated body, which is made of complicated carbon structures which absorbs oxygen from the air and when burned emits carbon dioxide, water, heat and produces a red light.

When the time process is reversed shifting from entropic to syntropic we would expect carbon dioxide, water, heat and red light frequencies to be absorbed. This would leave the complementary radiation to red which is green. If we look around we will notice that this syntropic process of green color really exists. This is the chlorophyll process, in the green leaves of plants which absorb carbon dioxide, water and heat. The chlorophyll process is therefore the dual process to the entropic one of combustion.

Studying and determining the laws of combustion in our laboratories can therefore allow us to account for the dual property of chlorophyll.

It is interesting to note that consciousness, the will and human personality, are processes which are oriented towards the future, moved by finalities and not causes. We can therefore state that psychical phenomena, our will and personality can generally be considered syntropic phenomena. For this reason they cannot be studied exhaustively using the experimental approach. It is also interesting to note that actions such as impulsive and emotional reactions which are caused by something that happened in the past are also those in which the activity of consciousness is reduced.

What makes life different is the presence of syntropic qualities: finalities, goals, and attractors. Now as we consider causality the essence of the entropic world, it is natural to consider finality the essence of the syntropic world. It is therefore possible to say that the essence of life is the final causes, the attractors. Living means tending to attractors.

The law of life is not the law of mechanical causes; this is the law of non-life, the law of death, the law of entropy; the law which dominates life is the law of finalities, the law of syntropy. But how are these attractors experienced in human life? When a man is attracted by money we say he loves money. The attraction towards a goal is felt as love.

We now see that the fundamental essence of life is love. I am not trying to be sentimental; I am just describing results which have been logically deducted from premises which are sure. The law of life is not the law of hate, the law of force, or the law of mechanical causes; this is the law of non-life, the law of death, the law of entropy.

The law which dominates life is the law of cooperation towards goals which are always higher, and this is true also for the lowest forms of life.

In humans this law takes the form of love, since for humans living means loving, and it is important to note that these scientific results can have great consequences at all levels, particularly on the social level, which is now so confused.

The law of life is therefore the law of love and differentiation. It does not move towards leveling and conforming, but towards higher forms of differentiation. Each living being, whether modest or famous, has its mission, its finalities, which, in the general economy of the universe, are important, great and beautiful.

Today we see printed in the great book of nature - that Galileo said, is written in mathematical characters - the same law of love that is found in the sacred texts of major religions.

ULISSE DI CORPO

I want to thank you all for your participation.

I am a psychologist and statistician. I was born in Rome, Italy, on January 26, 1959. I have always been exposed to very diverse cultures and religions and this brought me to reject anything which was dogmatic. My first vision of the universe was materialistic, but as a consequence of a year abroad, when I was at high school, I experienced a strong existential crisis, coupled with strong feelings of depression and anxiety. My materialistic vision did not help me to understand what was happening. I returned home hoping to go back to my old certainties, but my parents had just split in a very bad way and my materialistic vision continued to crumble.

Suddenly on the 19th of April 1977 I had the insight that consciousness and feelings require a property which is symmetrical and complementary to physical energy. This had a tremendous impact on me, since it took the form of a "Vital Needs Theory". This new vision helped me to explained the origin of depression and anxiety which just vanished.

Although I was gifted in mathematics and physics I enrolled in the faculty of psychology, where I was soon disappointed by the materialistic approach. I asked an astrophysicist to be my tutor, and my final dissertation considered the implications of this symmetrical energy in psychology. I then enrolled in a PhD in Statistics where the Dean recognized in my work the Unitary Theory of Luigi Fantappiè.

Fantappiè's publications were unavailable and I went on developing this theory on my own until I met Antonella Vannini in 2001 who provided impressive experimental evidence to this theory.

One major difference of my work is that I start from the dual-time solution of Einstein's energy/momentum/mass equation, whereas Fantappiè starts from the dual-time solution of the D'Alembert operator and consequently expects syntropic phenomena not only in quantum mechanics but also in electromagnetism.

We all link the Energy-Mass equation ($E=mc^2$) to Albert Einstein, but this equation was first published in 1890 by Oliver Heaviside, then in 1900 by Henri Poincare and in 1903 by the Italian Olinto De Pretto, who registered it at the *Regio Instituto di Scienze* and then published it in a paper together with the senator and astronomer Giovanni Schiaparelli.

It seems that the Energy-Mass equation reached Einstein through his father Hermann who was the owner of the "Privilegiata Impresa Elettrica Einstein", working in the development of street lighting in Verona together with Olinto De Pretto.

The $E=mc^2$ equation had a major problem, it did not take into account motion, the momentum, which is also a form of energy. Einstein solved the problem by adding the momentum, and publishing in 1905 the extended equation in his Special Relativity: the energy/momentum/mass equation.

The energy/momentum/mass equation is a double order equation

$$E^2 = m^2 c^4 + p^2 c^2$$

*Where **E** is energy, **m** is mass, **c** the constant speed of light and **p** the momentum*

and has two solutions for energy: a positive time solution, which describes energy that diverges from the past to the future, and a negative time solution, which describes energy that diverges backward from the future into the past. Since we move forward in time, the backward in time diverging energy turns into a converging force.

But, energy flowing backward in time was considered impossible and Einstein suggested to remove the momentum from the equation and go back to the $E=mc^2$, which always has only a forward in time solution. He could do this since the speed of physical bodies is practically nil compared to the speed of light.

Everything worked fine until 1924 when Wolfgang Pauli discovered that the spin of subatomic particles (which is a momentum) nears the speed of light. Consequently, quantum mechanics requires the use of the extended energy/momentum/mass equation, with its problematic backward in time solution! The first equation that combines Special Relativity and Quantum Mechanics dates back to 1926 and was formulated by Klein and Gordon. This equation has two solutions: a backward in time (advanced waves) and a forward in time (delayed waves). The advanced waves solution was rejected, since it implies retrocausality, which was considered impossible.

The second equation was formulated in 1928 by Paul Dirac. Dirac who tried to solve the paradox of the backward in time solution, but found the electron and the neg-electron (now named positron) that propagates backward in time.

Positrons were observed experimentally in 1932 and shortly after Pauli wrote an essay with the famous psychologist Carl Gustav Jung. Starting from the dual-time solution of the fundamental equations he posits that we live in a supercausal world, with causes acting from the past and synchronicities acting from the future.

But in 1933 Heisenberg, who had a strong charismatic personality and a leading position in the institutions and academia, declared the backward in time solution impossible.

The concept of energy comes from the observation that physical systems possess a quantity that can be turned into force. This magnitude can take many different forms: heat, mass, electromagnetism, potential energy, kinetic energy, nuclear and chemical.

However, modern science has not yet explained what energy is.

Richard Feynman, Nobel Prize for physics, says:

"It is important to realize that in physics today we have no knowledge of what energy is ... There is a fact, though, or if you want a law that governs all natural phenomena. There is no exception to this law. The law is called 'energy conservation' and states that the amount of energy does not change in the transformations it undergoes. This is an abstract idea, a mathematical principle that says that if there is an amount of energy, this remains constant. We can calculate the amount of energy and after any processing if we calculate again the amount of energy, the result is always the same."[10]

This is the first law of thermodynamics: *"Energy cannot be created or destroyed, but only transformed"*.

As we have seen, in Einstein's energy/momentum/mass equation ($E^2=m^2c^4+p^2c^2$) energy is double order and has two time solutions. But, since the future is for us invisible, we can say that two perfectly balanced realities exist: a visible and an invisible one. These two realities are united by the same energy and the same equation.

We can write:

$$E_{total} = E_{visible} + E_{invisible}$$

Total energy is the sum of visible and invisible energy

The visible reality expands and is governed by the law of entropy, whereas the invisible reality contracts and is governed by the law of syntropy.

We can also write:

$$E_{total} = E_{entropic} + E_{syntropic}$$

The first law of thermodynamics states that energy is a constant, it cannot be created or destroyed, but only transformed. We can consequently replace energy with the number 1 and write:

$$1 = Entropy + Syntropy$$

$$Entropy = 1 - Syntropy$$

$$Syntropy = 1 - Entropy$$

These equations show that entropy and syntropy are complementary parts of the same unity.

The definition of syntropy is therefore profoundly different from that of negentropy, which is defined as the negative of entropy:

$$negentropy = -entropy$$

This has incredible consequences since it implies that life and reality is the unity of these two opposite, but complementary realities.

The counter position life/entropy is continually debated by biologists and physicists. Schrödinger (Nobel prize for physics), answering the question about what allows life to contrast entropy, replied that life feeds on an energy which has symmetrical properties to those of physical energy.[11]

Albert Szent-Györgyi, Nobel prize for physiology and discoverer of vitamin C, used the word "*syntropy*" to describe the energy which is complementary to entropy.[12]

Life always shows the tendency to reduce entropy and increase syntropy. When it fails, entropy increases and the system goes towards degradation, suffering and death.

The energy/momentum/mass equation implies three types of time:

- *Causal time*, is expected in expanding systems, such as our universe, and it is governed by the properties of the positive time solution. In expanding systems entropy prevails, causes always precede effects and time moves forward, from the past to the future. Since entropy prevails, no advanced effects are possible, such as light waves moving backward in time or radio signals being received before they are broadcasted.
- *Retrocausal time*, is expected in contracting systems, such as black-holes, and it is governed by the properties of the negative time solution. In contracting systems retrocausality prevails, effects always precede causes and time moves backward, from the future to the past. In these systems no "delayed waves" are possible and this is the reason why no light is emitted by black-holes.
- *Supercausal times*, is expected in systems in which diverging and converging forces are balanced. Atoms are an example. Consequently, in these systems causality and retrocausality coexist and time is unitary: past, present and future coexist. This would be the reason why quantum mechanics is so different and always implies two levels, such as the particle and wave manifestation (causality/retrocausality).

This classification of time recalls the ancient Greek division in: kronos, kairos and aion.

- *Kronos* describes causal time, which is familiar to us,

made of absolute moments which flow from the past to the future.
- *Kairos* describes retrocausal time. According to Pythagoras, kairos is at the basis of intuitions, the ability to feel the future and to choose the most advantageous options.
- *Aion* describes supercausal time, in which past, present and future coexist. The time of quantum mechanics, of the sub-atomic world.

Syntropy and entropy coexist at the quantum level of matter, the Aion level, and at this level life can originate.

A question naturally arises: how do the properties of syntropy flow from the quantum level of matter to the macroscopic level of our physical reality, which is governed by the law of entropy, transforming inorganic matter into organic matter?

In 1925 Wolfgang Pauli provided the answer. He discovered in water molecules the hydrogen bonding.

Hydrogen atoms in water molecules share an intermediate position between the sub-atomic level (*Aion*) and the molecular level (*Kronos*), and provide a bridge that allows the properties of syntropy to flow from the quantum to the macro level.

Hydrogen bonds make water different from all other

liquids, increasing its attractive forces (syntropy), which are ten times more powerful than the van der Waals forces that hold together other liquids, with behaviour that are in fact symmetrical to those of other liquid molecules. Consequently water shows anomalous properties such as, when it freezes, it expands and becomes less dense. Other liquid's concentrate, solidify, and become more dense. In liquids the process of solidification starts from the bottom, whereas in water exactly the opposite happens.

The hypotheses that I put forward is that life originates at the quantum level, since at this level syntropy is available, and that life structures rapidly grow into the macroscopic level, governed by the opposite law of entropy. In order to survive the destructive effects of entropy, life needs to acquire syntropy from the quantum level and water provides the mechanism.

The law of syntropy implies the reformulation of the principles of thermodynamics:

- *Principle of conservation of energy*: energy can neither be created nor destroyed, but can only be transformed.
- *Principle of entropy*: in expanding systems energy is released, increasing homogeneity. Entropy is the magnitude with which the amount of energy that is lost into the environment is measured.
- *Principle of heat death*: in isolated systems placed in expanding systems (such as in our expanding universe)

entropy is irreversible, energy dispersion cannot decrease.
- *Principle of syntropy*: in converging systems energy is absorbed, increasing differentiation and complexity. Syntropy is the magnitude with which energy concentration, the increase in differentiation and complexity are measured.
- *Principle of heat concentration*: in isolated systems placed in converging systems syntropy is irreversible, energy concentration cannot decrease.

The perfectly balanced and complementary nature of entropy and syntropy also imply that systems, physical or biological, vibrate between entropy and syntropy. These vibrations take the form of pulsations, such as the heart beats, breathing and dynamic processes of expansion and contraction that characterize all living beings, and take the form of waves in physical phenomena such as light, sound and the wave nature of quantum mechanics.

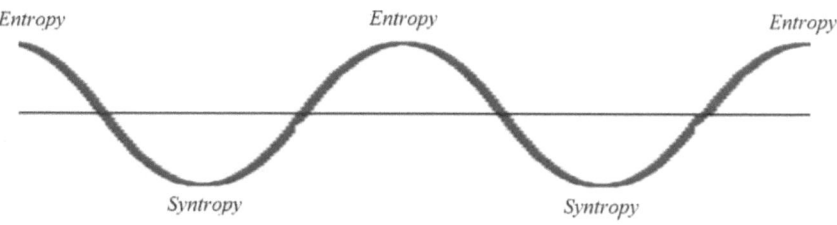

It is important to note that the concentration of energy cannot take place infinitely. When the limit is reached the process reverses and entropy takes over releasing energy and matter. In turn, the release of energy cannot be infinite, when the limit is reached the process reverses and syntropy prevails concentrating energy and matter.

This process activates an exchange of energy and matter with the environment: syntropy absorbs and organizes, entropy releases and destroys.

This continuous exchange is evident in metabolism in the form of:

- *anabolism* (syntropy) which absorbs energy and leads to the formation of complex biomolecules from simpler ones and nutrients;
- *catabolism* (entropy) that decomposes complex biomolecules in structurally simpler ones releasing energy in chemical (ATP) or thermal form.

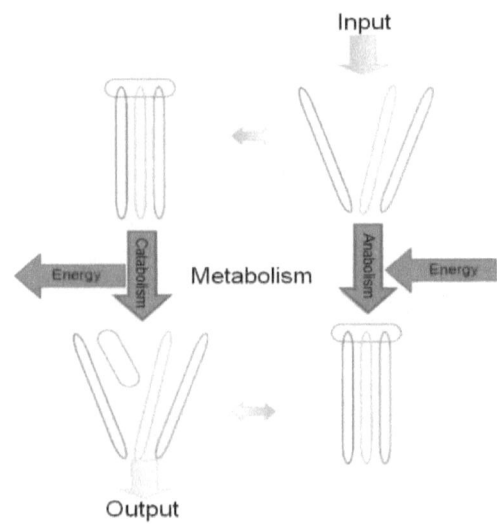

Since at our level entropy prevails metabolism requires an extra input of energy, which is generally provided by the energy of the Sun.

Metabolism is not an example of perpetual motion, since it will inevitably degrade into entropy. Instead, an example of perpetual motion is provided by the atom.

It is well known that an electric charge that undergoes acceleration, such as changes in velocity and direction as it happens with an electron, will emit electromagnetic radiation, losing energy in the process. A revolving electron should transform the atom into a miniature radio station, the energy output of which would be at the cost of the potential energy of the electron which would spiral into the nucleus and then the atom would collapse.

Well, this does not happen! We see the atom going on forever. Several explanations have been put forward, but they are all unsatisfactory.

The interplay of entropic and syntropic phases suggests that the atom is a vibrating system which alternates diverging and converging phases. In the diverging phase entropy prevails, whereas in the converging phase syntropy prevails rebalancing the effects of entropy.

The hydrogen atom vibrates at 10^{14} times a second. When it expands it releases a quanta of energy, when it contracts it absorbs a quanta of energy. This is the reason why energy is quantized, since it can be emitted only in the diverging phases and absorbed only in the converging phases.

In the diverging phase time flows forward, whereas in the converging phase time flows backward. For us, observing the atom from outside at a greater scale, the atom seems suspended in a unitary time in which past, present and future coexist.

A similar model was suggested by Einstein at the cosmological level. The endless vibration between expansion and contraction phases led Einstein in the 1920s to theorized a cyclical universe that endlessly shifts between Big Bang (expansion) and Big Crunch (contraction). During the Big Bang phase the universe expands until gravitational forces do not cause matter and energy to collapse. During the Big Crunch phase the universe contracts until diverging forces do not cause matter and energy to explode again in another Big Bang.

The term "Big Bang" was coined by Fred Hoyle during a BBC radio broadcast in March 1949. The first formulation of the theory of the Big Bang, by Lemaître, dates back to 1927, but was generally accepted only in 1964, when most scientists were convinced that experimental data confirmed that an event like the Big Bang took place.

Georges Lemaître, a Belgian Catholic priest and physicist, developed the equations of the Big Bang and suggested that the distancing of the nebulae was due to the expansion of the cosmos. He observed a proportionality between distance and spectral shift, now known as Hubble's law.

Edwin Hubble and Milton Humason showed that the distance of galaxies is proportional to their redshift, the shift of light towards lower frequencies. This happens usually when the light source moves away from the observer or when the observer moves away from the source. More specifically, it is called "red shift" when, in observing the spectrum of light emitted from galaxies, quasars, or distant supernovae, it appears shifted to lower frequencies when compared with the spectrum of closer corresponding objects. Since the red color is the lowest frequency in visible light, the phenomenon received the name redshift, even though it is used in connection with any frequency, including radio frequency radiations.

The redshift phenomenon indicates that galaxies are moving away from each other, and more generally that the Universe is in a phase of expansion. Redshift measurements show that galaxies and star clusters move away from a common point in space: the more distant they are from this point, the higher is their speed. Since the distance between galaxy clusters is increasing, it is possible to deduce, by going back in time, densities and temperatures increasingly higher until a point is reached where values tend towards infinite and the physical laws of the forward in time energy solution are no longer valid.

In cosmology, the Big Crunch is a hypothesis on the fate of the universe. This hypothesis is exactly symmetrical to the Big Bang and maintains that the universe will stop expanding and begin collapsing on itself.

According to the Big Crunch hypothesis the mutual gravitational attraction of all the matter of the universe will eventually cause the universe to contract. The strength of the gravitational force will stop the universe from expanding and the universe will collapse back on itself. While the early universe was highly uniform, a contracting universe will become increasingly diversified and complex. Matter will start to collapse into black holes, which will then coalesce producing unified black holes and eventually a Big Crunch singularity.

The cyclic theory proposes that the universe could collapse to the state where it began and then initiate another Big Bang, so in this way the universe would last forever, going through endless phases of expansion and contraction.

But, recent evidence, to be precise the observation of distant supernova, has led to the speculation that the expansion of the universe is not being slowed down by gravity but rather accelerating.

In 1998 the measurement of the light from distant exploding stars led to the conclusion that the universe is expanding at an accelerating rate. The observation of the redshift-luminosity of supernovae suggests that supernovae are spreading apart faster as the universe ages. According to these observations the universe appears to be expanding at an increasing rate. These observations contradict the hypothesis of the Big Crunch.

In the attempt to explain these observations physicists have introduced the idea of dark energy, dark fluid or phantom energy. The most important property of dark energy would be that it has a negative pressure which is distributed relatively homogeneously in space, a kind of anti-gravitational force which is driving the galaxies apart. This mysterious anti-gravitational force is considered to be a cosmological constant or vacuum energy which will lead the universe to expand exponentially. However, to this day no one actually knows what dark energy is, or where it comes from.

On the contrary the cyclic hypothesis suggests that the observed increase in the rate of expansion of the universe is not the effect of dark energy or any mysterious anti-gravitational force, but rather the effect of time slowing down: the acceleration is an illusion which is caused by time itself gradually slowing down.

In 1934 Richard Tolman rejected the cyclic model of the universe as it is incompatible with the second law of thermodynamics, which states that entropy can only increase. This implies that successive cycles of Big Bang and Big Crunch must be longer and wider than earlier ones, since entropy can only increase.

Einstein's cyclic hypothesis does not imply longer cycles since syntropy, during the converging phase of the Big Crunch, compensates entropy. When the maximum expansion or maximum cohesion is reached time reverses, giving rise to the opposite process. The universe moves back and forth in time. During the expansion phase, time flows forward, whereas during the contraction phase time flows backward. In the cyclic universe causality and retrocausality, entropy and syntropy, coexist and interact constantly.

Syntropy implies retrocausality. However, in the laboratories of physics it seems impossible to perform experiments on retrocausality since all the time-symmetric models lead to predictions identical with those of conventional models.[13] For this reason it is impossible to distinguish between time-symmetric results and conventional results.

In his transactional interpretations of quantum mechanics, John Cramer states that:

"Nature, in a very subtle way, may be engaging in backward in time handshaking. But the use of this mechanism is not available to experimental investigators even at the microscopic level. The completed transaction erases all advanced effects, so that no advanced wave signaling is possible. The future can affect the past only very indirectly, by offering possibilities for transactions."[14]

But, something special happens with gravity. We continually experience gravity, but do we know what gravity is? Can we cause it?

The negative time solution of energy suggests that gravity is a force that diverges backward in time. For us moving forward in time, it turns into a converging and attractive force, invisible since it originates from the future.

Can we test that gravity depends on the backward in time energy solution?

We know that forward in time energy cannot exceed the speed of light, whereas backward in time energy must always propagate at speeds beyond the speed of light, producing instantaneous effects. By measuring the speed of propagation of gravity we should therefore be able to test this backward in time hypothesis. If it is a manifestation of the backward in time energy solution its propagation must be instantaneous, otherwise it must not propagate at a speed greater than that of light.

Is it possible to perform such measurements?

The answer was provided by Tom van Flandern, an American astronomer specialized in celestial mechanics. Van Flandern noted that when measuring gravity no aberration is observed, and this poses the speed of propagation of gravity at 10^{10} times the speed of light.[15,16,17]

With light aberration it is due to its limited speed. For example, light from the Sun takes about 500 seconds to reach the Earth. So, when we look at the Sun we see it where it was 500 seconds before. This difference amounts to about 20 arc seconds, a large amount for astronomers. Sunlight hits the Earth from a slightly shifted angle and this is called aberration. If the speed of propagation of gravity is limited we would expect to observe gravity aberration. We should observe gravity coming from the position that the Sun occupied when gravity had left the Sun. But experiments show that there are no detectable delays in the propagation of gravity from the Sun to Earth. The direction of the gravitational pull of the Sun is exactly where the Sun is at the moment, not in a previous position, and this proves that the propagation speed of gravity is infinite.

Van Flandern notes that gravity has some special

properties. One of these is that its effect on a body is independent of its mass and that bodies fall in a gravitational field with the same acceleration, regardless of whether they are heavy or light. Another property is the infinite extension of the gravitational force. The extension cannot be infinite when forces propagate forward in time, at a finite speed. The other curious property of gravity is its action and instantaneous propagation, which can only be explained if we accept that gravity is a force that diverges backward in time.

As I have already said, the hypotheses that I have put forward is that life originates at the quantum level, since at this level syntropy is available. Life rapidly grows into the macroscopic level, governed by the opposite law of entropy. In order to survive the destructive effects of entropy, living systems need to acquire syntropy and to constantly reduce entropy.

The vital needs theory starts from this assumption. Living systems constantly struggle with entropy and, in order to survive, several material conditions must be met, such as drinking, food, shelter, and also several intangible conditions must be met, such as the need for meaning and the need for love. These conditions are vital, since when they are not met the living system dies.

When a vital need is met only partially an alarm bell is triggered. When we need water thirst is felt, when we need food hunger is experienced. The same applies for intangible needs, for example, depression is the alarm bell which informs that the vital need for meaning is not met and anxiety is the alarm bell which tells that the vital need for love is unsatisfied.

Let us see the three vital needs which arise from the interaction of life with entropy.

The first vital need is commonly known as ***material needs***. In order to combat the dissipative effects of entropy, living systems must acquire energy from the outside world, protect themselves from the dissipative effects of entropy and eliminate the remnants of the destruction of structures by entropy. These conditions include acquiring energy from the outside world through food and reducing the dissipation of energy with a shelter and clothing; disposing off wastes caused by entropy, and follow rules of hygiene and sanitation. Satisfying material needs leads to a state characterized by the absence of suffering. The partial satisfaction, however, is signaled by hunger, thirst and diseases. The total dissatisfaction leads to death.

The second vital need is commonly referred to as the ***need for love***. The satisfaction of material needs does not stop entropy from destroying the structures of living systems. For example, cells die and must be replaced. To repair the damages caused by entropy, living systems must draw on the regenerative properties of syntropy that allow to create order, regenerate structures and increase the level of organization. They must, therefore, acquire syntropy. In human beings this function is performed by the autonomic nervous system that supports vital functions. Since syntropy acts as an absorber and concentrator of energy the acquisition of syntropy is felt as feelings of warmth associated with wellbeing, in the thorax area where the autonomic nervous system is located. These feelings of warmth and wellbeing coincide with what is usually named love; the lack of syntropy is felt as a feeling of void and emptiness in the thorax area associated with suffering and distress, usually named anxiety and anguish. Briefly the need to acquire syntropy is experienced as need for love. When this need is not satisfied, feelings of void and suffering are felt. When this need is totally dissatisfied living systems are not able to sustain the regenerative processes and entropy takes over, leading the system to death.

The third vital need is commonly referred to as the ***need for meaning***.

In order to meet material needs we produce maps of the environment. These maps give rise to the identity conflict.

Entropy has expanded the physical universe towards infinite, whereas syntropy concentrates consciousness in extremely limited spaces. Consequently, when we compare ourselves with the infinity of the universe, we discover to be equal to zero.

On one side we feel we exist, on the other side we are aware to be equal to zero. These two opposite considerations *"to be, or not to be"* cannot coexist.

The identity conflict can be written in the following way:

$$\frac{I}{Universe} = 0$$

When I confront myself with the universe I am equal to nothing, zero

The universe corresponds to entropy whereas I corresponds to syntropy.

The identity conflict is characterized by feelings of nothingness and of being meaningless, by lack of energy, existential crises and depression. These feelings are generally perceived in the form of tensions in the head and generally come together with anxiety and anguish. To be equal to zero is equivalent to death, which is incompatible with our feelings of existence.

We must therefore solve the conflict. Most people try to increase their meaning through wealth, power, achievements, judgment of others, a purpose, ideologies and religions.

$$\frac{I+judgment+wealth+popularity+power\;\cdots}{Universe} = 0$$

But whatever we put at the numerator compared with an infinite universe continues to be equal to zero.

The identity conflict can be solved only thanks to the theorem of love:

$$\frac{I \times \cancel{Universe}}{\cancel{Universe}} = I$$

When I unite with the universe, compared with the universe, I am always I

It is important to note that the multiplication "x" corresponds to the cohesive properties of love.

Only when we love, we can remove "Universe" from the numerator and denominator and the equation becomes I = I.

This demonstrates that when we unite with the universe through love, the identity conflict between being and non-being (I = 0) is solved and turns into a confirmation of our identity: I = I.

In other words, love solves the identity conflict and provides a meaning to existence. It also solves the conflict between syntropy and entropy and allows the transition from duality to non-duality.

The Theorem of Love shows that the final aim of life is love.

ANTONELLA VANNINI

I also want to thank the organizer for this beautiful opportunity.

I am a psychotherapist and hypnotherapist. I was born in Rome, Italy, on September 14, 1972 and discovered the Unitary Theory in 2001 when I met Ulisse and I chose to enroll in cognitive psychology where I studied in depth the Unitary Theory.

My first dissertation *"Entropy and Syntropy, from mechanical to life sciences"* was published in the NeuroQuantology Journal, my master dissertation titled *"Entropy and Syntropy: causality and retrocausality in psychology"* was published in the Syntropy Journal and my PhD dissertation *"A syntropic model of consciousness"* has been published by ICRL, Princeton, in a book titled *"Syntropy, the spirit of love."*

During my PhD I conducted several experiments in order to study the retrocausal hypothesis which stems from the Unitary Theory.

The idea of retrocausality has been always rejected since in the laboratories of physics it seems impossible to perform experiments that support the validity of this hypothesis.

On the contrary in the laboratories of psychology, biology and life sciences it is easy to perform experiments that demonstrate the hypothesis of retrocausality.

The theory of syntropy postulates that syntropy is the energy of life. Consequently systems that support vital functions should show retrocausal activations. In humans, the autonomic nervous system supports vital functions. It is, therefore, assumed that its parameters, such as heart rate and skin conductance, should show retrocausal activations.

Pre-stimuli activations seem to play a key role in the survival and welfare of all living systems. Robert Rosen, for example, coined the expression *Anticipatory Systems*. He was amazed by the amount of anticipatory behavior observed at all levels of the organization of living systems that behave as true anticipatory systems. Systems in which the present state changes according to future states, violate the law of classical causality according to which changes depend solely on past or present causes. Scientists have tried to explain this behavior with theories and models that exclude any possibility of anticipation. Without exception, all the theories and biological models are classical in the sense that they only seek causes in the past or present.[18]

Various experiments show the existence of anticipatory pre-stimuli reactions of skin conductance and heart rate, for example:

A study performed in 1997 by Dean Radin monitored heart rate, skin conductance and fingertip blood volume in subjects who were shown a blank screen for five seconds and a randomly selected calm or emotional picture for the following three seconds.[19] Radin found significant differences, in the autonomic parameters preceding the exposure to emotional versus calm pictures.

In 2003 Spottiswoode and May replicated Radin's experiments, adding controls to exclude artifacts and alternative explanations.[20] Results showed an increase in skin conductance 2-3 seconds before emotional stimuli are presented.

Similar results have been obtained by other authors, using various parameters of the autonomic nervous system, for example: McCraty, Atkinson and Bradley[21], Radin and Schlitz[22], May, Paulinyi and Vassy[23].

Daryl Bem, psychology professor at the Cornell University, studied retrocausality using well known experimental designs in a "time-reverse" pattern.

In his 2011 article *"Feeling the Future: Experimental Evidence for Anomalous Retroactive Influence on Cognition and Affect"*[24], Bem describes 9 well-established psychological effects in which the usual sequence of events was reversed, so that the individual's responses were obtained before rather than after the stimulus events occurred.

For example in a typical priming experiment the subject is asked to judge if the image is positive (pleasant) or negative (unpleasant), pressing a button as quickly as possible. The response time is registered. Just before the image a "positive" or "negative" word is briefly shown. This word is named "prime". Subjects tend to respond more quickly when the prime is congruent with the following image (both positive or negative), whereas the reaction times become longer when they are not congruent (one is positive and the other one is negative). In retro-priming experiments, results show the classical priming effect with faster reaction times when the prime shown after the response is congruent with the image.

During my PhD in cognitive psychology, I conducted four experiments using heart rate measurements in order to study Fantappiè's retrocausal hypothesis. A detailed description of these experiments is available in the book *"Retrocausality: experiments and theory"*. [25]

Each experimental trial was divided into 3 phases.

Presentation phase: 4 colors are presented one after the other on the screen of the computer: blue, green, red and yellow. Each color is shown for exactly 4 seconds. The subject is asked to look at the colors, and during the presentation the heart frequency is measured at fixed intervals of 1 second. For each color 4 measurements of the heart frequency are saved: one each second.

Choice phase: at the end of the presentation phase, an image with the 4 color bars is shown in order to allow the subject to choose the color which he thinks the computer will select. In other words, the subject is asked to guess the color which the computer will select.

Random selection of the target: as soon as the subject chooses a color the computer selects the target color, using a random process, and shows the selected color full-screen on the computer.

Phase 1				Phase 2	Phase 3
blue	green	red	yellow		TARGET

Target is the color selected and shown by the computer in the third phase.

The *hypothesis* was the following: in the presence of a retrocausal effect differences should be observed between heart rates measured in phase 1 according to the target randomly selected in phase 3.

Trials were repeated 100 times for each subject and

subjects were supervised by the experimenter only during the first trial and left alone for the remaining 99 trials. Consequently, the first trial was not considered in the data analyses.

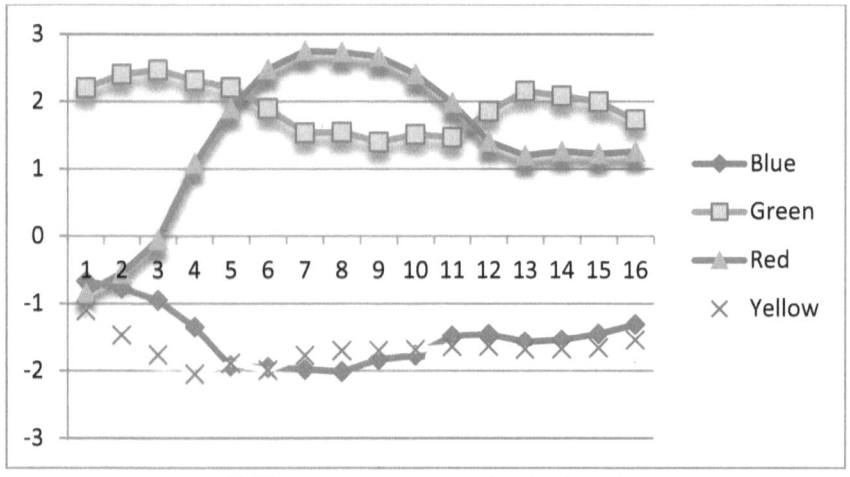

*Differences among mean values in phase 1
according to the target color randomly selected in phase 3
(the data is relative to one subject)*

In the absence of retrocausality heart rate frequencies should vary around the 0.00 line. But, results show that they diverge from the 0.00 line, when grouped according to the target color.

Each subject shows a characteristic pattern in these retrocausal heart rate reactions. Some subjects increase heart rates when the target color is blue and reduce it when the target is green. Others show a pattern that is exactly the opposite.

Although in all experiments a strong retrocausal effect is

observed, subjects did not show the ability to guess correctly in phase 2. On the whole, the target is guessed slightly more than 25% of the times. Once out of four is exactly what it is expected by chance. In other words, our rational side, the head, seems unable to access the pre-stimuli information that the autonomic nervous system, the heart, already feels.

As Rainer Maria Rilker says: "*The future enters into us, in order to transform itself in us, long before it happens.*"

But, we must learn to listen to it.

These feelings of the heart seem to be fundamental in decision-making and in order to orient the person towards the future.

The neurologist Antonio Damasio, studying patients affected by decision making deficits, noted that specific lesions of the prefrontal cortex, especially in those sectors which integrate signals arriving from the body, lead to an absence, or the imperfect perception of feelings and to a behaviour which he described as "*short-sighted toward the future*". Damasio suggested a somatic markers hypothesis, according to which feelings constitute a part of the decision making process.

Damasio describes somatic markers in the following way:

> "*a negative feeling is felt in the stomach before the negative outcome of a decision. Because this feeling is relative to the body, I used the technical name somatic state; and because it marks an image, the word marker.*"[26]

Somatic markers can be measured as reactions of the autonomic nervous system using the parameters of skin conductance and heart rate.

Somatic markers have been tested by Bechara through the Iowa Gambling Task.[27] Participants were asked to choose cards from four different decks. Two decks ensured modest winnings but limited losses, while the other two, provided immediate gains but led to severe losses.

Whereas normal participants learned to choose from the first two decks, patients with lesions of the prefrontal cortex chose the last two decks, even when the implicit rule became explicit. During the experiment three types of reactions of the autonomic nervous system emerged, two after the choice and one before. The first two reactions appear after the gratification or punishment due to the loss; the third response manifests before choosing from an unlucky deck. The anticipated response shows only in normal subjects with no lesions of the prefrontal cortex.

Damasio interprets the anticipatory reaction of skin conductance as an effect of learning.

A very simple example, which has puzzled many of us, is provided by the strange strategy cats use when they want to jump on a table:

They are unable to see what is on the table, but they smell the food and want to get on it. They first start circling the table till they choose a spot. Then they start assessing the jump moving in a slow motion their back. But what are they assessing, since it is impossible for them to see the top of the table? They cannot rely on any rational information for their assessment. And still, when they jump, they land perfectly in the most narrow spots!

According to the retrocausal hypothesis they engage a

game with their feelings and future, assessing in this way the outcome. They try infinite invisible jumps and feel the results. When the feeling is of certainty, they know that that is the solution and they jump.

A similar situation was described by the mathematician Henri Poincaré. He noticed that when faced with a new mathematical problem he began using the rational approach of the conscious mind that allows to become aware of the elements of the problem. But, since the options are infinite and it would take infinite lives to evaluate them all, some other type of process leads to the correct option. This process selects the solution, among all the infinite possibilities in an unconscious way. Poincaré named it intuition or inspiration and noticed that it is always coupled with a feeling of certainty and beauty:

"The useful combinations are precisely the most beautiful, I mean those best able to charm this special sensibility that all mathematicians know, but of which the profane are so ignorant as often to be tempted to smile at it. What happens then? Among the great numbers of combinations blindly formed by the subliminal self, almost all are without interest and without utility; but just for that reason they are also without effect upon the aesthetic sensibility. Consciousness will never know them; only certain ones are harmonious, and, consequently, at once useful and beautiful. They will be capable of touching this special sensibility and which, once aroused, will call our attention to them, and thus give them occasion to become conscious. ... Thus it is this special aesthetic sensibility which plays the rôle of the

delicate sieve of which I spoke, and that sufficiently explains why the one lacking it will never be a real creator." [28]

Let us use another metaphor. In a bucket filled with water no complexity arises by effect of chance. Water is distributed according to the law of entropy and a state of maximum entropy, of maximum stillness, reigns.

When an attractor is introduced, for example we unplug the bucket, water starts flowing in a preferential direction and organizes itself in those ways that lead to the attractor, without diverging from it.

Syntropy is the great unifying attractor of life, which is also love. When it starts operating it works just like the plug in a sink. There is a preferential direction and this direction is highlighted by feelings of love, warmth, wellbeing and beauty, whereas the direction that goes towards an increase of entropy and death is signaled by void, suffering and anguish.

Most people experience presentiments. In many cases these feelings of the future can be beneficial.

For example, visceral feelings have been reported to save lives: *"suddenly I felt a sense of cold associated with danger and I shouted: no - no!"* Feelings of terror can lead to choose differently and avoid death.[29] In a study on commuters trains accidents William Cox[30] found that when a train has an accident the number of passengers is considerably lower than expected. He made controls (departure time, day of the week, weather conditions), but he always found that when a train has an accident the number of passengers on board is lower than expected. Visceral feelings seem to inform about the accident, causing conditions that lead not to board the train. This seems to be the case also in airplane accidents. When boarding (that is after checked-in), around 2% of the passengers feel ill and in many cases they are reported not to board the plane.

Visceral feelings alert about the future, but they use a language which is archaic. Animals use this archaic language, that we call "instinct", and this allows them to feel, with days in advance, natural disasters.

The first report dates back to 373 B.C., when animals, including rats, snakes and weasels, left en masse the Greek city of Elice few days before a devastating earthquake. Animals panicked, dogs started barking and whining for no apparent reason.

In China, where the invisible energy of life is taken seriously into account, these strange behaviour are used as alarm bells. For example, in 1975 people of Haicheng, a city with one million people, were ordered to flee their homes. A few days later a magnitude 7.3 earthquake destroyed the city. If the abnormal behaviour of animals had not been taken seriously, more than 150,000 people would have died.

In most languages we find the distinction between feelings and emotions, but we often confuse these two words. In the Unitary Theory feelings are syntropic, linked to attractors and to the future, whereas emotions are entropic, linked to past experiences.

Feelings provide the information in order to converge towards the attractor. Since syntropy is energy that converges, feelings of warmth and of wellbeing in the autonomic nervous system area tell that we are on the right path. On the contrary feelings of void, chill and suffering tell that we are on the wrong path.

Feelings are usually hidden under our emotions and the chatter of our mind. Consequently, in order to decide well we need to learn how to feel and how to calm the chatter of the mind.

A very effective way is provided by Zen meditation. During Zen meditation participants cannot react to stimuli, but they can only observe them. Practicing Zen meditation we discover that thoughts wait for the reaction of the heart. When the heart reacts it provides energy to the thought which becomes stronger. When we don't react the thought dissolves. The heart decides when to react and when to be silent; the mind can only adjust to the will of the heart. We are the heart. Our will is in the heart. In this way the scepter of command moves from the head to the heart and the mind becomes silent. The importance of silence can be found in many traditions. Shared silence helps to calm the chatter of the mind and to focus on feelings.

The heart is our will ! This statement provides an explanation to some strange findings.

Benjamin Libet[31], researcher in the physiology department of the University of California, San Francisco, conducted some pioneering experiments in the field of will. His experiments measured the readiness potentials and the activation of volition.

Results show that muscles start acting before the will is activated, before volition takes place. The interpretation was that we are machines which react automatically and that free will is just an illusion of our mind.

The Unitary Theory, on the contrary, suggests a totally different interpretation. The heart is syntropic, consequently the will must work according to advanced potentials. That means that we expect the activation of muscles before volition takes place.

EPILOGUE

The Unitary Theory can be applied in the most diverse fields: from physics, biology, psychology, economics, social sciences, arts, teleology and theology.

Some brief examples will be now provided.

- The role of intuitions

An example on how intuitions are linked to the future and to wealth has been offered by Steve Jobs, the founder of Apple Computer. This example also shows the difficulties in the harmonization of the visible and invisible

Jobs was born in 1955 and was the founder of the most successful company in human history.

Jobs had ventured in India, from where he returned with a changed vision of life. He used to repeat that:

"People in the Indian countryside do not use their intellect like we do, but they use intuitions. Intuitions are very powerful, more powerful than the intellect."

In 1976, in a friend's house, he saw the circuit board of a

computer and had the intuition of people using personal computers. Jobs had learned in India that intuitions point to the future. Going against the opinion of others, who considered personal computers the stuff for few "crazy" minds, he asked Steve Wozniak to develop a prototype, which he named Apple I. He managed to sell a few hundreds of them. The success of Apple I led to a more advanced model for ordinary people: the Apple II. Jobs had an artist mind, not a technical one. His insights were mainly based on aesthetics and minimalism, which combined together made Apple II a commercial success.

What helped Steve Jobs to be intuitive was his frugal life that kept him away from entropy. He was vegan, practiced Zen meditation and liked to spend time in nature. He was intuitive, but also irrational. He used to argue continually with the "rationalists" and with John Sculley, manager that he had brought to the direction of Apple Computer. In 1985 the conflict became so severe that the board decided to fire Jobs from Apple Computer, the company that he had founded. Apple Computer went on living on the products that Jobs had designed, but after a few years the decline started.

In the mid-nineties Apple Computer was on the brink of bankruptcy and on December 21, 1996, the board asked Jobs to return as the personal adviser to the president. Jobs agreed. He asked a salary of one dollar a year and the guarantee that his insights, albeit crazy, had to be accepted without any condition. In a few months he revolutionized the products and on September 16, 1997 he became CEO ad interim. In less than a year he resuscitated Apple Computer and turned it in the company with the biggest profits of any company and the largest market value.

How did he manage?

"Do not let the noise of others' opinions drown your own inner voice. And most important, have the courage to follow your heart and intuition. They somehow already know what you truly want to become. Everything else is secondary."

Although Jobs was able to generate immense fortunes, money was not his property, but a tool for reaching an end. The ability to intuit was his wealth, his creativity, genius and innovation.

Einstein believed that: *"The intuitive mind is a sacred gift and the rational mind is his faithful servant. But we have created a society that honors the servant and has forgotten the gift."*

Jobs' attention was in the heart and he had no fear of death:

"Almost everything, all external expectations, all pride, all fear of embarrassment or failure, these things just fall away in the face of death, leaving only what is truly important. Remembering that you are going to die is the best way I know to avoid the trap of thinking you

have something to lose. You are already naked. There is no reason not to follow your heart."

Jobs often said that his mission, his attractor, was a computer that could be held in a hand. He died a few months after the presentation of the iPad, the computer that can be held in one hand. His life testifies that wealth comes from the invisible world, through insights and intuitions that reduce entropy and anticipate the future.

- Attractor in biology: the example of Syntropic Agriculture

When we try to explain the intelligence and order of genetic information solely as a result of past causes, we are faced with logical contradictions and paradoxes, since the processes of random mutations is a product of entropy, and can only lead to an increase in entropy, thereby preventing the formation of species. With life we witness an incredible convergence of biological structures towards common designs, despite individual differences.

For example, we can definitely indicate different races, such as Europeans, Asians, Africans, but there is something that unites all of them, and that makes them all human beings.

Considering only the influx of the past it is impossible to explain why individuals converge towards the same designs, and their stability in time. Attractors which retroact from the future can instead explain all of this.

Experiments devised by Rupert Sheldrake can shed some light. For example, when individuals of a species learn to solve a task the knowledge is spread in an invisible and immaterial way to all the other individuals of the same species.

Attractors behave as relays. Individual information arrives to the attractor and it is assessed. When it is advantageous, for example when an individual solves a task, the information is selected by the attractor and it is disseminated to all the other individuals. Attractors provide a bridge between individuals, a shared knowledge. Members of the same attractor, such as animals of the same species, are able to share knowledge in an invisible way, without any physical mean.

These experiments have been replicated and are very simple. For example, when rats in a laboratory learn to solve a task which provides a reward, all the other rats of the same species, all around the globe, show the tendency to solve the same task more quickly. The same effect can be observed whenever attractors are at play. For example, when a new crystal is devised the process of crystallization becomes faster all over the World.

The Unitary Theory suggests that attractors receive

information and experiences from individuals, select that which is advantageous and redistribute it using the retrocausal channel of feelings.

This process changes information into in-formation. Intelligent information, which provides solutions, designs and projects. The verb "to inform" comes from the Latin "in-formare", that means "to give a form".

Aristotle believed that "in-formation" is a primitive fundamental activity of energy and matter. In-formation does not have an immediate meaning, such as the word "knowledge", but rather it encompasses a modality that provides form.

Once a form takes place in the attractor, it can be expressed throughout all the individuals linked to it. The autonomic nervous system plays a key role since it connects individuals to the attractor and in this way to all the other individuals and life forms, gathering in-formation.

Despite the incredible amount of intelligence that in-formation requires, it is present at all the levels of the organization of life. It does not depend on the conscious mind and free will, but it is a quality of the unconscious mind. The autonomic nervous system, i.e. the unconscious mind, behaves like a mechanic who consults the book of the manufacturer to perform repairs and maintain the system as close as possible to the project. The project is not mechanical and instructions are written with the ink of the heart.

Since we can access in-formation thanks to feelings and intuitions, feelings and intuitions are at the basis of any syntropic activity. For example, syntropic agriculture has been developed by Ernst Götsch in Brazil. Götsch just feels what the soil and plants need. Using this intuitive approach he is able to transform degraded soils into above average yields and at the same time increase biodiversity. He is able to turn deserts into forests, making the soil rich in nutrients, for high-quality organic agriculture.

After years of intensive use of pesticides and fertilizers, soils have become arid and agriculture production is starting to decrease. It is therefore vital to shift towards an agriculture which is able to regenerate the soil. Syntropic agriculture seems to be capable of doing this, since it is based on the law of syntropy.

Götsch has been contacted by multinationals in the agro-business field, interested to know how to use this approach at the industrial level. But, the problem is training. How can people be trained to become intuitive and feel what the soil and plants need? How can we establish the connection with the attractor and feel the in-formation?

An example is provided by artists. The most divine musicians feel the music they are playing. They fall in a state of trance which connects them directly with their heart. The most outstanding chefs don't follow recipes, but they feel how to combine ingredients. The connection with the attractor is established at the unconscious level, where the mind is silenced, and the heart can express itself. This becomes easier when we learn how to fall in states of trance.

- Unity in diversity: the self-organizing Universe and life on Earth

Attractors bring parts together. The unity of our Self is strengthened when we are converging towards the attractor. When, on the contrary, we have no attractor cohesion diminishes, the chatter of the mind increases and our personality shatters.

Converging is therapeutic since it brings together our parts and makes them cooperate. The evolutionary paleontologist Teilhard de Chardin noticed that the incredible stability of species is given by the fact that they converge and he advocated the idea that life is guided by attractors, and evolves according to a hierarchy of attractors, till the ultimate unifying attractor, the Omega point, is reached. Since attractors reinforce the Self, they increase individualization and differentiation, nonetheless they also lead towards unity. It seems a contradiction, but unity and diversity go together.

The theme of attraction has been the focus of Teilhard's research:

"Reduced to its essence the problem of life can be expressed like this: accepting the two principles of conservation of energy and entropy, how can they assimilate without contradiction, a third universal law (which is expressed by biology), that of the organization of energy? ... the situation becomes clear when we consider, at the basis of cosmology, the existence of a sort of anti-entropy."

Teilhard formulated the hypothesis of a converging energy, similar to syntropy:

"...not just one kind of energy, but two different energies; two energies which cannot transform directly one into the other, because they operate at different levels ... The behaviour of these two energies are so completely different and their manifestations so completely irreducible that we might believe they belong to two completely independent ways of explaining the world. And yet, as the one and the other, are in the same universe, and evolve at the same time, there must be a secret relationship."

Attractors, Omega point, syntropy, purpose and mission are synonyms. This can cause confusion. Mission or purpose are typically used for individuals, "Omega point" for the source of syntropy.

Lately the biologist Rupert Sheldrake has coined the expression "morphic fields" to indicate attractors. Morphic fields establish and organize invisible webs of relations that form an important part of our lives. A kinship which facilitate collaboration, cooperation and synergy.

- Beauty and aesthetics in scientific creations

Most people would think that science is simply mechanical, but it is not. It is not merely a question of applying rules, of making the most combinations possible according to certain fixed laws. The true work of the inventor consists in creating, and the rules which guide this process are felt rather than formulated.

Beauty and aesthetics usually guide scientific achievements. Inventing does not consist in making new combinations with entities which are already known, but to discover new entities and meaningful combinations. The possibilities are infinite and a whole lifetime would not suffice to examine them all.

Scientific discoveries are guided by intuitions, coupled by strong feeling of warmth in the heart area: aesthetics and beauty, and mystical experiences. Most striking at first is the sudden illumination of a solution, by a feeling of absolute certitude. Those scientists who do not feel this beauty are unable to invent.

These aesthetic feelings guide the creation, and belong to

sensibility. Now, what elements are capable of developing an aesthetic feeling? They are those harmoniously disposed so that the mind without effort can embrace their totality while realizing the details. This harmony is at once a satisfaction of our aesthetic needs and an aid to the mind. And at the same time it makes us foresee a new law. The useful combinations are the most beautiful, those best able to charm this special sensibility that all inventors know. It is this special aesthetic sensibility that sufficiently explains why the one lacking it will never be able to create.

- The role of death

The invisible side of reality is the most important for life. We can feel it, but we cannot see it. It can be associated to what people usually name soul or spirit.

We are incarnated souls. The body is subject to the effect of entropy and dies, whereas the soul is subject to the effect of syntropy and gradually evolves towards the final attractor of love.

The soul is immortal, does not die, but needs to do physical experiences in order to learn and evolve. When we are incarnated we need to provide a meaning to our existence and this leads us to attach to whatever provides a meaning. At a certain point our evolution stops and death is needed to go on in our path.

We constantly vibrate between the visible and invisible, between life and death. Death is not the end, but just the transition towards the invisible, whereas birth is the transition towards the visible. Whereas birth is the transition into the world dominated by entropy, death is the transition towards the world dominated by syntropy. Death is part of our process of evolution and we should not fear it.

- The future of humanity

We are witnessing one of the most difficult moments in human history. Wars everywhere, the risk of a Third World War, pollution, criminality, mental illnesses which have tripled in the last 15 years, drug abuse, families disintegrating, indebted nations and skyrocketing taxes. We are now in a period dominated by entropy and suffering, and the majority of the population believes that there is no way out.

On the contrary, if life is sustained by syntropy, the Earth system which sustains life should show retrocausal activations. This means that an apocalypse in the future would break the retrocausal chain of events and life would be impossible anyway down to the present and the past.

Consequently, just the fact that we exist is the proof that

we will continue to evolve towards the attractor, which at the end will be love and cooperation. This path will not be easy, since people don't want to change. And if people don't want to change suffering will be inevitable.

As Rainer Maria Rilker says: *"The future enters into us, in order to transform itself in us, long before it happens."*

But, we must learn to listen to it.

NOTES

1. Wave phenomena are represented by differential equations with second order derivatives of the hyperbolic type, whereas in order to describe the phenomena studied by classical mechanics and by optics equations with first order derivatives are used (Jacobi equations) or the equivalent ordinary differential equations (canonical mechanical equations). This implies that whereas in classical mechanics we can distinguish trajectories of entities with their own individuality, in wave mechanics the presence of equations with partial derivatives of an hyperbolic order greater than one, leads to phenomena which are not localized, with the change of time, in a limited area (just think of the space occupied by a particle).
2. Schrödinger's wave equation takes the Hamiltonian function H, which characterizes the system in classical mechanics and measures the total energy relative to its space coordinates and to the momentums, and writing that the wave equation (which describes with the square of its modulus the probabilistic density) has a variation in time (a first derivative relative to time, using the mathematical language) which is proportional, for a constant factor, to an expression which is obtained applying to the same function a linear differential operator, which is obtained from the Hamiltonian function replacing the momentums with the derivatives of the corresponding variables, changed using a constant factor. Since the Hamiltonian function is squared for the momentums, a linear expression of the second derivatives is obtained referring only to the spatial variables, and a term which contains the unknown function y (which is relative to the potential), and a last term in which the first derivative is relative to time. In the case of a single particle with the space

coordinates x, y, z, Schrödinger's wave equation is a linear differential equation of the second order, which contains the first derivative relative to time, and the second derivatives of the space variables are always parabolic (since the particle is a H term which is expressed by a polynomial of the second order in the momentums), of the same kind of the equation that governs the conduction of heat in solid matter.

3. The most important properties of the second derivative equation which was initially obtained by Dirac are obtained from the characteristic cone, which is determined by the second order terms of the equation. As we have seen these terms are obtained applying the D'Alembert operator to the unknown function, and consequently the characteristic cone is always real, matching the chronotope which, with the vertex in the assigned event, divides the events from the future to the past ones and from those which can be concomitant, according to Special Relativity. Consequently from this structure of the characteristic cone the value of the unknown function y of the assigned event (that is to say in the point of the chronotope with coordinated x,y,z,t), at least in the case of the events which we have previously determined, can depend only on the values of y and eventually on the terms of the equation (which represents the density of the distribution of the sources of the wave propagation) known from the past events, whereas the value of the y point and of the known term can influence only the values that y acquires in the field of the future events. In other words the field dependence of the solutions of the event which has been considered is attributed only to the past events, whereas the field influence to the future events, whereas events outside of the chronotope cannot influence or be influenced by the event. For those who are less familiar with the four dimensional representation of the chronotope, it is sufficient to say that the past events, that is events which fall within the boundaries of the cone, are given for each instant before the one we are considering t, by the points

within a sphere with its centre in the points x,y,z with a radius which decreases with the speed of light, till it reaches zero in the instant t, whereas the future events are given, for each instant following t by the points of a sphere, with the same centre, with a radius which increases with the speed of light, starting from the zero value at the instant t.

4. This can be clearly stated following another path; if we just consider that in wave phenomena the partial derivatives equations which describe them need to be of the hyperbolic type, and need to satisfy special relativity, the values of the solutions of a point x,y,z at an instant t, for any phenomena which we have caused, must be the consequence of values within the converging sphere towards the point at the speed of light (past events according to special relativity) and can effect only those points within the sphere which diverges from the same point, with the same speed (future events according to special relativity), otherwise if an element outside these two regions could affect or be affected from the event, the action between the two events should propagate at speeds which are greater than the speed of light, which according to special relativity is impossible.
5. H. Poincaré, Electricité et optiqtee, 2.e éd., Paris, 1901
6. W. Ritz, Recherches critigues sur l'électrodinantique générale, Ann de physique, 8 s., t. 13, 1908, p. 145
7. G. Giorgi, Sulla sufficienza delle equazioni differenziali della fisica matentatica, Rend. Lincei, s. Ga, vol. VIII, 1928. Per un'ampia bibliografia sull'argomento, cfr. A. Cabras, Sulla teoria balistica della luce, Mem. Lincei, s. 6a, vol. III, f. 6°, 1929.
8. Starting from the hypothesis that the wave always starts from a source, with a density measured by the second known member of the equation; this solution is obtained in each point as the sum (integral) of the infinitesimal contributes (potentials) due to the sources, distributed in the single elements of the volume, in previous instants (to that which is being considered) at a certain

time, that is needed for the wave to diverge at the speed of light c, from the volume element where the source is situated at the point considered;

9. The definition of cause which we give here coincides with the definition that Galileo gave: "A cause is that which when present is followed by an effect and when removed the effect disappears."
10. Feynman R (1965), The Feynman Lectures on Physics, California Institute of Technology, 1965, 3.
11. Schrödinger E. (1944), What is life?
12. Szent-Gyorgyi A (1977), Drive in Living Matter to Perfect Itself, Synthesis 1977, 1(1): 14-26.
13. Wheeler J.A. and Feynman R.P. (1949), Classical Electrodynamics in Terms of Direct Interparticle Action, Reviews of Modern Physics 21 (3): 425–433.
14. Cramer J.G. (1986),The Transactional Interpretation of Quantum Mechanics, Reviews of Modern Physics, Vol. 58: 647-688.
15. Van Flander T. (1996), Possible New Properties of Gravity, Astrophysics and Space Science 244:249-261.
16. Van Flander T. (1998), The Speed of Gravity What the Experiments Say, Physics Letters A 250:1-11.
17. Van Flandern T. and Vigier J.P. (1999), The Speed of Gravity – Repeal of the Speed Limit, Foundations of Physics 32:1031-1068.
18. Rosen R (1985) Anticipatory Systems, Pergamon Press, USA 1985.
19. Radin DI (1997), Unconscious perception of future emotions: An experiment in presentiment, Journal of Scientific Exploration, 11(2): 163-180.
20. Spottiswoode P (2003) e May E, Skin Conductance Prestimulus Response: Analyses, Artifacts and a Pilot Study, Journal of Scientific Exploration, 2003, 17(4): 617-641.
21. McCratly R (2004), Atkinson M and Bradely RT,

Electrophysiological Evidence of Intuition: Part 1, Journal of Alternative and Complementary Medicine; 2004, 10(1): 133-143.
22. Radin DI (2005) e Schlitz MJ, Gut feelings, intuition, and emotions: An exploratory study, Journal of Alternative and Complementary Medicine, 2005, 11(4): 85-91.
23. May EC (2005), Paulinyi T e Vassy Z, Anomalous Anticipatory Skin Conductance Response to Acoustic Stimuli: Experimental Results and Speculation about a Mechanism, The Journal of Alternative and Complementary Medicine. August 2005, 11(4): 695-702.
24. Bem D (2011), Feeling the future: Experimental evidence for anomalous retroactive influences on cognition and affect, Journal of Personality and Social Psychology, Jan 31, 2011.
25. Vannini A e Di Corpo U, Retrocausality: experiments and theory, Kindle Edition, ASIN: B005JIN51O (2011).
26. Damasio AR (1994), Descarte's Error. Emotion, Reason, and the Human Brain, Putnam Publishing, 1994.
27. Bechara A (1997), Damasio H, Tranel D and Damasio AR (1997) Deciding Advantageously before Knowing the Advantageous Strategy, Science, 1997 (275): 1293.
28. Henri Poincaré, Mathematical Creation, from Science et méthode, 1908.
29. In Battle, Hunches Prove to be Valuable, The New York Times on July 28, 2009,
30. Cox, W.E. (1956), "Precognition: An analysis," Journal of the American Society for Psychical Research, 1956(50): 99-109.
31. Libet, B (1985), "Unconscious cerebral initiative and the role of conscious will in voluntary action," The Behavioral and Brain Sciences 8: 529-566.
32. Mill J.S. (1843), A System of Logic, University of Toronto Press, 1843.

www.ingramcontent.com/pod-product-compliance
Lightning Source LLC
Chambersburg PA
CBHW020927180526
45163CB00007B/2910